Mel Bay Presents

A Bass Player's Guide to Jamming

by Carl Yaffey

CD Contents

1. Introduction

Tunes plus Dialog:
2. Hot Corn, Cold Corn
3. Roll in My Sweet Baby's Arms
4. Nine Pound Hammer
5. John Henry
6. Two Little Boys
7. Columbus Stockade Blues
8. Down the Road
9. Little Maggie
10. John Hardy

Tunes in Stereo:
11. Hot Corn, Cold Corn
12. Roll in My Sweet Baby's Arms
13. Nine Pound Hammer
14. John Henry
15. Two Little Boys
16. Columbus Stockade Blues
17. Down the Road
18. Little Maggie
19. John Hardy

1 2 3 4 5 6 7 8 9 0

© 2006 BY MEL BAY PUBLICATIONS, INC., PACIFIC, MO 63069.
ALL RIGHTS RESERVED. INTERNATIONAL COPYRIGHT SECURED. B.M.I. MADE AND PRINTED IN U.S.A.
No part of this publication may be reproduced in whole or in part, or stored in a retrieval system, or transmitted in any form
or by any means, electronic, mechanical, photocopy, recording, or otherwise, without written permission of the publisher.

Visit us on the Web at www.melbay.com — E-mail us at email@melbay.com

Contents

About the Author 2
Preface 3
Acknowledgements 4
Anatomy of a Jam Session 5
Bluegrass Rhythm 6
The 1-5 Method 7
Chords and Keys for Jamming 8
Participating in a Jam Session 9
Jamming Etiquette 11
About the CD 12
Appendix 1 – Finding the Notes on an Upright Bass 13
Appendix 2 - Some Typical Guitar Chords 14
Appendix 3 - Resources 15
Appendix 4 - Some Rules for Jamming 16
Appendix 5 - CD Tunes 18
Appendix 6 – Using an Electric Bass For Bluegrass 27

ABOUT THE AUTHOR

Carl Yaffey, a multi-instrumentalist (banjo, guitar, bass, mandolin), has been playing music on and off since 1962 (the "folkie" days). Way back then in Norfolk, VA, he played banjo and guitar in several Kingston Trio "clone" groups. After moving to Columbus, OH in 1966 to accept a job as a programmer with Bell Laboratories, he played bass in a country band and several Rock and Roll "oldies" bands. In between, Carl has played old-time and bluegrass banjo and bass with several different groups. Carl now teaches music, plays with Grassahol and the Timbre Wolves. And, he jams whenever he can.

Carl is the author of "A Banjo Player's Guide to Jamming", Mel Bay (MB99708BCD) "A Mandolin Player's Guide to Jamming", Mel Bay (MB20818BCD) "A Flatpicker's Guide to Jamming", Mel bay (MB20820BCD) and co-author of "A Fiddle player's Guide to Jamming", Mel bay (MB20889BCD).

PREFACE

You're a beginner or intermediate upright or electric bass player; you are familiar with a few tunes such as "Cripple Creek", "Old Joe Clark", and "Boil Them Cabbage Down". You might have learned the basics of playing, but you may not have played with other people much or at all. You're at your favorite festival, and you've just come up on a **HOT** jam session. You'd love to join in, **BUT** you're afraid that:

a) You won't know the tunes.
b) You won't know what to do or how to do it.
c) You'll be embarrassed or choke.
d) They'll ignore you and pretend you're not there.
e) Someone will take your bass out of your hands and fling it into the woods.
f) They'll all get up and leave.
g) They'll laugh.
h) Something even more horrible might happen.

Been there, done that! Well, not (e), but certainly most of the rest. It doesn't have to be like that! My task is to give you some tools and tricks so you can sit in on most any jam session and come out alive and grinning. It's easier than you think. Here's the secret: **LEARN TO PLAY ALONG WITH THE CHORDS IN COMMON KEYS.** Ha! I hear you saying, "I know about chords – I just don't always know what to play along with them and when to play." Never fear, I will cover that and more. Let's get started!

ACKNOWLEDGEMENTS

Thanks to the many musicians I have met and played with over the years. Pete, Chas, Paul, the many Mikes, the various Johns, Nick, Polina, Andi, Janet, Dave, and Jim just to name a few. You've given me an excellent education. I want to give special thanks to the musicians that jam at the Bluegrass Music Store in Columbus, OH – Dan, Al, Brink, Bud, Merilee, Tom, Rich, Tyler, Jerry, John, and all the rest. A welcoming jam if ever there was one.

Thanks to my wonderful wife, Debbie, for never inhibiting my desire to play music and for always applauding me when I do.

ANATOMY OF A JAM SESSION

A jam session is an informal, varied group of musicians playing together. Typically, members of the group come and go, have different abilities, and often don't all know each other. There may be two or twenty. There may be one of each instrument, or ten guitars or six banjos. The jam may be hot, lukewarm, or very, very cold.

Jam sessions spring up at festivals, flea markets, music stores, and campgrounds. Pretty much anywhere there's a bunch of musicians. Some are open and some are closed. If you see four musicians sitting with their knees almost touching, their heads down, and playing extremely well together, it's probably a closed jam. They might be members of a band practicing. Sit back and be the audience! But if there's a circle of musicians with some empty spaces between them and they seem to be relaxed and having fun, and one smiles and calls out to you to "come on in", it's definitely an open jam.

Then, there's everything in between! How do you make sure it's an open jam before you jump in? Well, you might not be able to tell from "outside". If it's not obviously closed, be brave and join in if there's space. Of course, you won't want to do the wrong thing and immediately feel unwanted! See "Jamming Etiquette" for the proper way to comport yourself in a jam session.

At any particular time, there's likely to be a leader. It may be someone who suddenly starts playing and/or singing a tune, or it may be someone who says, "Hey, let's play Rocky Top in G". In some jams, the leader changes often. In others, there may be a set leader.

In a bluegrass jam, as in a bluegrass band, the players take turns playing lead while the others play "backup" (hopefully). And, everyone plays backup during the singing (once again, hopefully!). Note that in an old-time jam, no one takes a lead break. The melody instruments all play the melody together. Folk jams are more singing oriented and most everyone will be playing backup or accompaniment.

As a bass player, you can join in using the "1-5 method" which I will describe.

BLUEGRASS RHYTHM

Let's talk about bluegrass rhythm. The first beat of a tune is called the "downbeat". The next beat is the "backbeat". These two beats alternate over and over in a song. A vast majority of bluegrass tunes will be in 4/4 time, with 4 beats in each measure: two downbeats, and two backbeats.

Each instrument in a typical bluegrass band or jam session will have a particular rhythm role associated with these beats as shown here:

INSTRUMENT	DOWNBEAT	BACKBEAT
Guitar	Pick bass note	Strum
Bass	Pick bass note	nothing
Banjo	Optionally pick a note	Simultaneously pick 3 quickly muted strings.
Mandolin	Optionally pick a note	Quick strum with quickly muted strings
Fiddle	nothing	Tap strings with the bow
Dobro	nothing	Quick strum with muted strings

As you can see, the bass plays on the downbeats.
Here is an example of this in the first measure of "Boil Them Cabbage Down":

BEAT	Downbeat	Backbeat	Downbeat	BACKBEAT
WORDS	boil	them	cab-	-bage
BASS	Play "G"		Play "D"	

The bass notes indicated, G and D, will be explained in the next section, "The 1-5 Method". Some tunes are in 3/4 time, "waltz time". Here are four measures of "Angel Band":

BEAT	D	B	D	B	D	B	D	B
WORDS	Lat-	-est	Sun	is	Sink-	ing	fast	my
BASS	Play "G"		Play "D"		Play "G"		Play "D"	

THE 1-5 METHOD

The 1-5 method is very simple. You play the "root" note of the chord on the first downbeat. You play the fifth note of the chord on the next downbeat. You alternate from there. What are the root and fifth notes of a chord? Let's look at a major scale. Here is the G major scale:

G	A	B	C	D	E	F#	G
1	2	3	4	5	6	7	8

The root note is the first note, G. The fifth note is the fifth note of the scale, D. So, if the chord being played is G, you would play a G and then a D. How about C? Here is the C major scale.

C	D	E	F	G	A	B	C
1	2	3	4	5	6	7	8

You play C and then G.

Ok, now that we know which notes to play, where on the instrument do we play them? If the chord is G and you play the open G string on the first downbeat and then play the open D string on the next downbeat, you are playing the simplest 1-5 possible. Notice that when you play the D note, you move directly across the neck to the next lowest string. So, to play 1-5 against a D chord, you could play the open D string on the first downbeat, and the open A string on the next.

This works fine for open strings. But what about chords where the root note is not on an open string? Lets look at C. The root note, C, is on the 3rd fret of the 3rd string on the electric bass, and on the 3rd position of the 3rd string of the upright bass. The fifth of the C chord is G. If you move from the 3rd fret/position of the 3rd string to the same fret/position of the 4th string, you will have a G note. There is another way to play 1-5. Instead of playing the 5 on the fret/position below the fret/position of the root, you can play it on the next higher string two frets/positions higher. Here is a diagram that shows several ways to play 1-5:

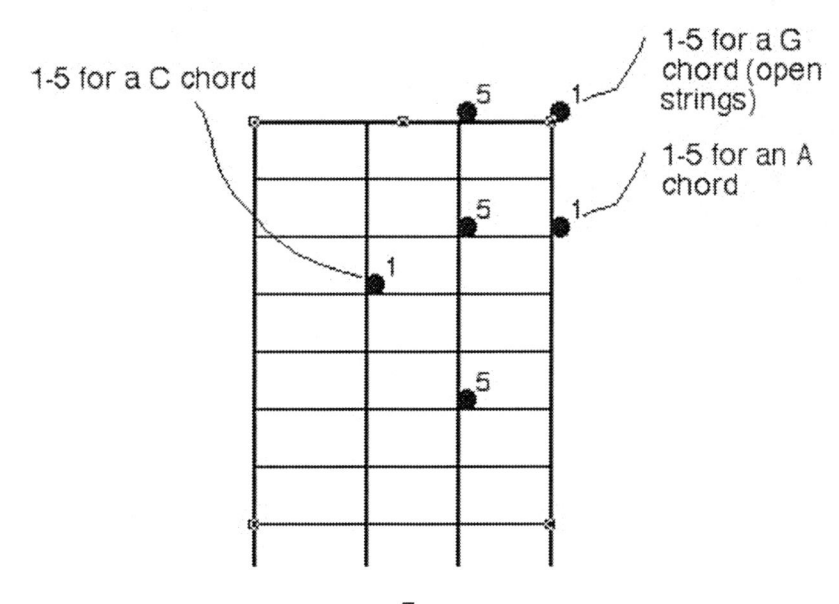

CHORDS AND KEYS FOR JAMMING

Tunes are played in a particular "key". Here's a simplified definition:

A key is a set of chords that sound good together.

Here are some typical keys and the chords that typically go with them:

KEY	CHORDS
G major	G, C, D7
A major	A, D, E7
C major	C, F, G7
D major	D, G, A7
E major	E, A, B7

Why these particular chords? There is a pattern here! To see it, let's examine the first key, G major. This is a key used for many bluegrass tunes. First, look at the notes that make up the G major scale:

G	A	B	C	D	E	F#	G
1	2	3	4	5	6	7	8

Note that the fourth note is C and the fifth note is D. That's the pattern for finding the major chords of any major key: 1, 4, 5. Let's try the pattern on the key of C major. The C major scale:

C	D	E	F	G	A	B	C
1	2	3	4	5	6	7	8

Sure enough, 1, 4, and 5 are C, F, G.

You'll note that in the table above, the third chord of a key has a "7" beside it. Often, the third chord has a fourth note added to it to make it a "seventh" chord. That note is the flatted 7th note of the scale. Seventh chords for guitar are shown in Appendix 2.

To play your bass along with these chords, you will use the "1-5 method".

This method depends on you knowing the notes on the strings of your bass. If you are just beginning to play upright bass, you might be having trouble with this. If so, flip over to Appendix 1 to learn how to find the notes on the strings.

PARTICIPATING IN A JAM SESSION

Congratulations! You can now play along with ANY common bluegrass key. Now let's find out how to do that in a jam session.

You'll need to know WHAT the key is, and WHICH chords are being played WHEN. Here are some ways:

1. Practice at home with CDs, tapes, tablature, and chord charts.
2. Watch a guitar player.
3. Listen.
4. Ask questions.

Practice at home with CDs, tapes, and chord charts

There is really no substitute for simply being very familiar with the music. If you want to join bluegrass jam sessions, listen to a BUNCH of bluegrass music. The CD included with this book presents nine familiar, traditional bluegrass tunes performed at a jam session with the jammers on one channel and a spoken description of what's happening on the other. Listen to the tracks and then try to play along. You can also skip to tracks 11-19 where the tunes are in stereo (no dialog). Attempt to figure out what key the tune is in. Note that the first and/or last chord played often indicates the key. Now, try to play along. You can simply play "root" notes at first since you're just trying to match the chords. If you have trouble, try listening to what the guitar player is doing. Finally, I've given you the key and chord progressions in the section "about the CD" so that you can check yourself.

Once you can successfully "join in" the jam session on the CD, get chord charts for some other tunes. These can be found in tablature, many songbooks, and on the Internet. Pick out a song you like, find a CD that includes it, play the CD and "read" along on the chord chart for the tune. Note what key the song is in. The chart or songbook page will indicate this. If not, remember that the first and/or last chord played usually indicates the key with the last melody note being the "tonic" or key note; i.e., in the key of C, the last melody note will often be C. The more songs you do this with, the better you'll get at hearing the changes without knowing them in advance.

Watch a guitar player

Watch a jam session that includes a guitar player. Assuming the guitar player is competent, watch his/her left hand. You will see WHEN the chords change and you will see WHICH chords are being played. This is an invaluable skill! If you need more help with recognizing chords, Appendix 2 presents some typical guitar chords, and there are some valuable resources for learning guitar chords listed in Appendix 3.

Listen

Pay attention to what is being played in a jam session. Listen closely to the sound as the chords change. Couple this with watching a guitar player. See if you can name the chords (silently to yourself!) as they change.

Ask questions

Don't be shy! After a tune is finished, ask a friendly-looking person a question or two. "What was that chord you played in the second part?"

JAMMING ETIQUETTE

Yes, there is a jamming etiquette. If you don't know it, it's not likely that anyone will correct you when you violate it – often they'll just grumble to their friends and probably not be very friendly to you later. Depending on how serious your violations were, you may find yourself labeled a "jam buster" and become unwelcome in the future. You might even be told to "get lost" right on the spot! I've even seen a jam close itself down and move to another place to get away from a rude player.

Some of this etiquette is common sense, and some is not. First of all, the jam may or may not be welcoming joiners. You should check out the body language. As I mentioned in the section "Anatomy of a Jam Session", if the musicians are huddled close together, and there's no room between them, it's probably a closed jam. If someone in the jam looks at you, smiles, and says, "Hey, sit down and join us," it's clearly an open jam. What about in between? If you're not sure, you could simply ask someone between tunes.

If you're too shy to do that try moving into the circle and QUIETLY playing along with the next tune. Notice the reaction of the players. Frowns? Smiles? Nothing? Frowns indicate either they don't want anyone joining or you are breaking a jamming rule. "Nothing" may also indicate this. But, if you don't think you're getting "go away" signals, keep QUIETLY playing. The important thing here is to not be rude. Playing louder than the lead player or the singers is rude and unmusical. Playing loudly between tunes is rude. Playing lots of notes and "showing off" during someone's lead is rude. I have listed some jamming rules in Appendix 4. They're primarily for bluegrass jams, but most apply to old-time and folk jams as well. If you follow these, you will not be regarded as rude. In fact, you'll be in great shape!

ABOUT THE CD

The CD[1] contains 9 tunes performed at a typical jam session. On the left channel is a group of musicians[2] singing and playing guitar, banjo, mandolin, and bass. The right channel contains some dialog about the tunes such as the name, the key, and the chord progression. If you shift your stereo's balance control (assuming you have one) all the way to the left, you will be able to play along without hearing the dialog. When you want to hear the group plus the dialog, shift the balance control back to center. I suggest that you try to play along with the group on the left channel at first. If you have trouble, listen to the right channel to hear the dialog. Each tune is charted in Appendix 5 where you will find the key, chord progression, and words. Try your best to play along with the CD at first without listening to the dialog or looking at Appendix 5.

Following the 9 tunes with dialog is another set of tracks containing the same tunes in stereo. The CD contains the following:

Track 1 Introduction
      Tunes with dialog:
Track 2 Hot Corn, Cold Corn
Track 3 Roll in My Sweet Baby's Arms
Track 4 Nine-Pound Hammer
Track 5 John Henry
Track 6 Two Little Boys
Track 7 Columbus Stockade Blues
Track 8 Down the Road
Track 9 Little Maggie
Track 10 John Hardy
Tunes in stereo:
Tracks 11 – 19

---

[1] The CD was recorded and mixed by John Sherman.

[2] Mark "Brink" Brinkman, banjo and vocals; Rich Baker, mandolin and vocals; Dan Cook, guitar and vocals; Carl Yaffey, bass.

APPENDIX 1 – Finding the Notes on an Upright Bass

An electric bass conveniently has frets where the notes are located. But an upright bass does not. How do you know where to place your fingers for the various notes? Many beginners mark "frets" or positions on their bass neck as a way to get started. Here's how:

Let's start with the first string, the G string. Pressing about 2 inches down from the nut will give an A♭ note. That's the equivalent of the first fret of an electric bass guitar. Check this with a tuner.

Now mark the edge of the fingerboard. You can do this with a pencil, or you can place a stick-on label or piece of tape there. Move your finger to a spot around 2 inches farther from the A♭ mark. Pressing there will give an A. When you're sure you have the position correct, mark the fingerboard.

Now, mark the fingerboard five more times, each mark around 2 inches farther along the fingerboard and check for the notes B♭, B, C, C♯, and finally D.

Use this same procedure for the E, A, and D strings:

The D string notes will be E♭, E, F, F♯, G, G♯, A.

The A string notes: B♭, B, C, C♯, D, D♯, E.

The E string notes: F, F♯, G, A♭, A, B♭, B.

APPENDIX 2 – Some Typical Guitar Chords

------- KEY OF G -------

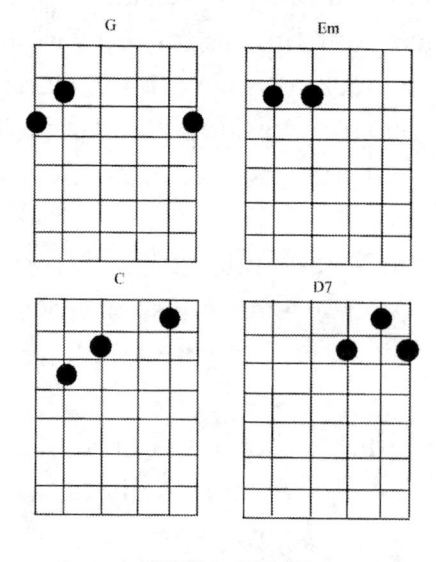

G   Em

C   D7

------- KEY OF C -------

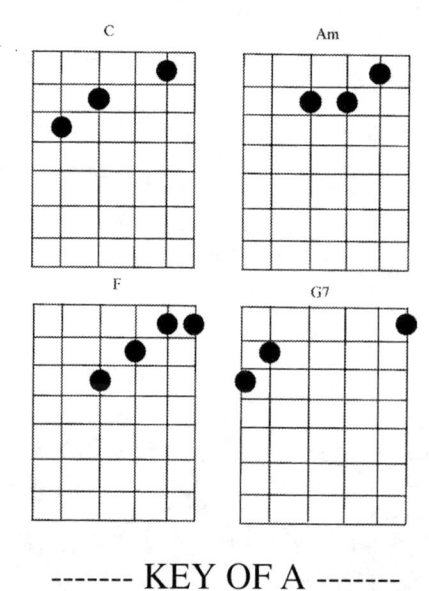

C   Am

F   G7

------- KEY OF A -------

A   F#m

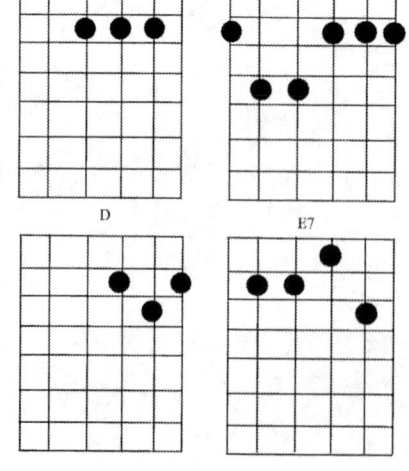

D   E7

APPENDIX 3 – Resources

1. "Bluegrass Bass" • Video • Mel Bay Publications (MB94928VX)
2. "Electric Bass" • Video • Mel Bay Publications (MB94924VX)
3. "String Bass Fundamentals" • Video • Mel Bay Publications (MB96773VX)
4. "Learn to Play Bluegrass Bass" • Mel Bay Publications (MB93638)
5. "Bluegrass Bass Favorites" • Mel Bay Publications (MB20314)
6. "You Can Teach Yourself Guitar Chords" • Mel Bay Publications (MB95120)
7. "Guide to Capo, Transposing, and the Nashville Numbering System" • Mel Bay Publications (MB98413)
8. "Instant Guitar Chord Finder" • Mel Bay Publications (MB94824)
10. "Guitar Chord Chart" • Mel Bay Publications (MB93322)

APPENDIX 4 - SOME RULES FOR JAMMING

TUNE YOUR INSTRUMENT. There are too many good, cheap electronic tuners around not to do this. Do it QUIETLY or away from the group.

LISTEN. If you can't hear the vocal or lead instrument, then YOU are playing too loud. Be VERY sensitive to volume levels. In a bluegrass jam, it is VERY IMPORTANT to NOT play loudly and NOT play lead when someone is singing or taking a lead break. Play backup!

INDICATE WHO THE NEXT SOLOIST WILL BE. You might be the leader on a tune. When handing off an instrumental break on, say, a fiddle tune, follow a pattern (for example, clockwise around a circle) so the next lead player knows it is his/her turn. If the intended lead player passes, he/she can indicate with a shake of their head. You can also indicate who's next by nodding at them. If you are not the leader, pay attention to who's next or who gets the "nod". Do NOT play out of turn!

WELCOME OTHERS. Open the circle if other players wish to join. Jam sessions cannot be too large if everyone is polite. You know how it feels to be excluded or unwelcome. Don't do that to others.

SHARE THE SELECTION. Open the choice of songs to individuals around the circle. Don't monopolize the jam by playing one song after another. An instrumentalist can suggest that a vocalist do a song to make things more interesting.

TRY NEW STUFF. Once in awhile, a participant in a jam may suggest an original tune, or one that is out of character with the jam, or a song not easily followed by the group. This is OK occasionally. Just don't overdue it.

LET OTHERS KNOW WHEN YOU'RE NOT JAMMING. Bands may sometimes be rehearsing and may need to exclude non-band members from joining in. If this is the case, an explanation would be appropriate.

DO NOT RAID. Don't interrupt an active jam by calling favorite musicians away to begin another jam. After a reasonable amount of time, any jam session will change its personnel.

KEEP STEADY RHYTHM. Errors in rhythm are the most difficult to overcome while keeping a group together. Avoid adding and dropping beats. If you're having trouble keeping the proper rhythm, keep your volume WAY DOWN.

DO NOT RUSH. Don't start a song too fast for the other players. Once everyone has had a turn on an instrumental, then someone can announce that the tempo is going to accelerate.

DO NOT "NOODLE" BETWEEN TUNES. If you want to practice or try out a lick you just heard, wait until you get away from the group or until you get home. It will help to have a small tape recorder with you to save a tune to practice against later. Don't play between tunes to show off. If you would like the group to play a given tune, speak up!

APPENDIX 5 – CD Tunes

Hot Corn, Cold Corn

Key of G

There are only two chords in this tune – G and D.

G
Hot corn, cold corn bring along a demijohn
D
Hot corn, cold corn bring along a demijohn
G
Hot corn, cold corn bring along a demijohn
D   G
Fare thee well, uncle Bill, see you in the morning, yes sir.

Well, upstairs downstairs, down in the kitchen
Upstairs downstairs, down in the kitchen
Upstairs downstairs, down in the kitchen
See Uncle Bill, he's a-raring and a-pitching, yes sir.

Old Aunt Peggy won't you fill 'em up again
Old Aunt Peggy won't you fill 'em up again
Old Aunt Peggy won't you fill 'em up again
Haven't had a drink since I don't know when, yes sir.

Yonder come the preacher and the children are a-crying
Yonder come the preacher and the children are a-crying
Yonder come the preacher and the children are a-crying
Chickens are a-hollering, toenails a-flying, yes sir.

Roll in My Sweet Baby's Arms

Key of G

Three chords G, C, and D. A very common set.

```
G
Well, I ain't gonna work on the railroad
                        D
I ain't gonna work on the farm
    G                   C
I'll lay around the shack 'till the mail train comes back
    D               G
I'll roll in my sweet baby's arms.
```

Chorus:
Roll in my sweet baby's arms
Roll in my sweet baby's arms
Lay around the shack 'till the mail train comes back and
Roll in my sweet baby's arms

Now where were you last Saturday night
While I was layin' in jail?
Walking the streets with another man
Wouldn't even go my bail

I know your parents don't like me
They drove me away from your door.
If I had my life to live over
I wouldn't go there anymore.

Nine Pound Hammer

Key of A

Three chords A, D, and E. Another very common set.
```
A                         D
```
This nine-pound hammer is just a little too heavy
```
        A E         A
```
Honey, for my size, Honey for my size.
```
                          D
```
Roll on buddy, don't you roll so slow
```
        A E               A
```
How can I roll when the wheels won't go.

It's a long way to Harlan it's a long way to Hazard
Just to get a little booze, Just to get a little booze.
Roll on buddy, don't you roll so slow
How can I roll when the wheels won't go.

Oh, this nine pound hammer that killed John Henry
Ain't gonna kill me, ain't gonna kill me.
Roll on buddy, don't you roll so slow
How can I roll when the wheels won't go.

There ain't one hammer in this tunnel
That can ring like mine, that can ring like mine.
Well it rings like silver, and it shines like gold,
Oh, it rings like silver, and it shines like gold.

When I'm long gone you make my tombstone
Out of number nine coal, out of number nine coal.
Roll on buddy, don't you roll so slow
How can I roll when the wheels won't go.

Going on the mointain, just to see my baby
And I ain't coming back, no I ain't coming back.
Roll on buddy, don't you roll so slow
How can I roll when the wheels won't go.

Well this nine-pound hammer is just a little too heavy
Honey, for my size, Honey for my size.
Roll on buddy, don't you roll so slow
How can I roll when the wheels won't go.

John Henry

Key of D

Two chords D, and A.

D
John Henry was a little baby boy, you could
                            A
Hold him in the palm of your hand. His
D
Papa cried out this lonesome farewell:

Johnny gonna be a steel driving man, Lord, Lord,
            A   D
Johnny gonna be a steel driving man.

John Henry went up on the mountain
His hammer was striking fire.
But the mountain was so tall, John Henry was so small
He laid down his hammer and he died, Lord, Lord,
Laid down his hammer and he died.

John Henry went into the tunnel
Had his captain by his side.
The last words that John Henry said:
Bring me a cool drink of water 'fore I die, Lord, Lord,
Cool drink of water 'fore I die.

John Henry had a little woman
And her name was Polly Ann.
John Henry took sick and he had to go to bed
Polly drove that steel like a man, Lord, Lord,
Polly drove that steel like a man.

Well you can talk about John Henry as much as you please
Say all that you can.
There's never been born in the United States
Never such a steel driving man, Lord, Lord,
Never such a steel driving man.

Two Little Boys

Key of A

Four chords A, B, D, and E.
A
Now two little boys had two little toys
     A7   D
Each had a wooden horse.
             A
Together they played one sunny day
B   E
Warriors both of course.
A
One little chap had a mishap
       A7   D
And broke off his horse's head.
          A
He wept for his toy then cried for joy
B   E   A
When he heard his brother say:

Chorus:
A   D   A
Do you think I could leave you crying
            A7   D
When there's room on my horse for you?
         A
Climb up here Jack and stop your crying
B   E
You'll mend up your horse with glue
A   D   A
When we grow up we'll both be soldiers
      A7   D
Our horses will not be toys
      A
And maybe you will remember
B   E   A
When we were two little boys.

Now long years had passed, war came at last
Bravely they marched away.
And the cannons roared loud, and in that wild crowd
Where wounded and dying Joe lay.
Then came a cry, a rider dashed by
Out from the ranks of blue.
He galloped away to where Joe lay
And he heard his brother say:

Chorus:
Do you think I could leave you dying
When there's room on my horse for you?
Climb up here Joe, we'll soon be flying
Among the ranks of the boys in blue.
Can't you see Jack I'm all a-tremble
It may be the flash and the noise,
Or it may be because I remember
When we were two little boys.

Columbus Stockade Blues

Key of G

Three chords G, C, and D.

```
G
Way down in Columbus Georgia,
    D          G
I'd rather be in Tennessee.

Way down in Columbus stockade,
    D                    G
My friends have turned their backs on me.
```

Chorus:
```
C            G
Go and leave me if you wish to,
 C              D
Never let it cross your mind.
         G
In your heart you love another,
D              G
Leave me darling I don't mind.
```

Last night as I lay sleeping
I dreamed I held you in my arms.
When I awoke I was mistaken
I was peeking through these bars.

Many a night with you I rambled
Many an hour with you I spent.
Thought I had your heart forever
Now I find it was only lent.

Go and leave me if you wish to,
Never let it cross your mind.
In your heart you love another,
Leave me darling I don't mind.

Down the Road

Key of G

Three chords G, Em, and D.

```
G                       Em
Down the road about a mile or two
G              D    G
Lives a little girl named Pearly Blue.
G              Em
About so high hair of brown
G              D    G
Prettiest thing boys in this town.
```

Chorus:
```
G                   Em
Down the road, down the road
G              D    G
Got a little pretty girl down the road
```

Now old man Flatt he owns the farm
From the hog lot to the barn,
From the barn to the rail
Makes his living by carrying the mail.

Now every day and Sunday too
I go to see my Pearly Blue.
Every time the rooster crows
See me headed down the road.

And every time I get the blues
I walk the soles right off my shoes.
Don't know why I love her so,
Gal of mine lives down the road.

Down the road about a mile or two
Lives a little girl named Pearly Blue.
About so high hair of brown
Prettiest girl now in this town.z

Little Maggie

Key of A

Three chords A, G, E

```
A                    G
Over yonder stands little Maggie
A         E    A
With a dram glass in her hand
                     G
She's drinking away her troubles
      A       E   A
And courting some other man.
```

Well flowers were made for blooming
And stars were made to shine,
Oh women were made for loving
Little Maggie was meant to be mine.

Lay down your last gold dollar
Lay down your gold watch and chain.
Little Maggie's gonna dance for daddy
Now listen to that mandolin ring.

Last time I saw little Maggie
She was sitting on the banks of the sea,
With a forty-four around her
And a banjo on her knee.

Go away, go away little Maggie
And do the best you can.
I'll find me another woman
You can get you a brand new man.

John Hardy

Key of A

Although this is done as an instrumental on the CD, I decided to give you the words in case you had the urge to sing it!

Three chords A, D, and E.

```
      D         A
John Hardy was a desperate little man,
      D              A
He carried two guns every day.
      D              A
He shot a man on the West Virginia line,
E
You should have seen John Hardy getting away, poor boy,
                         A
You should have seen John Hardy getting away.
```

He went on across to the East Stone Bridge
There he thought he'd be free.
Up steps the sheriff and he takes him by the arm
Saying: Johnny come along with me, poor boy,
Johnny come along with me.

He sent for his mama and his papa too
To come and go his bail.
But there weren't no bail on a murder charge
So they threw John Hardy back in jail, poor boy,
Threw John Hardy back in jail.

John Hardy had a pretty little girl
The dress that she wore was blue.
She came into the jailhouse hall
Saying: Johnny I'll be true to you, poor boy,
Johnny I'll be true to you.

APPENDIX 6 – Using an Electric Bass For Bluegrass

It's no secret that some folks do not like to hear an electric bass in a bluegrass tune. The probable reason is that they've not heard one played correctly. The mistake some EB players make is to let the notes "ring" and "boom". To fix this mute the notes right after making them. Your bass will then sound more like an upright bass. Here's how: if you play an open string note, you mute the note using your left hand on the backbeat. If you play a fretted note, you mute the note by lifting your fretting finger on the backbeat. The notes should not ring past the downbeat. It will also help to adjust your amplifier so that you are not too loud.

EXCELLENCE IN MUSIC

MEL BAY®

Since 1947

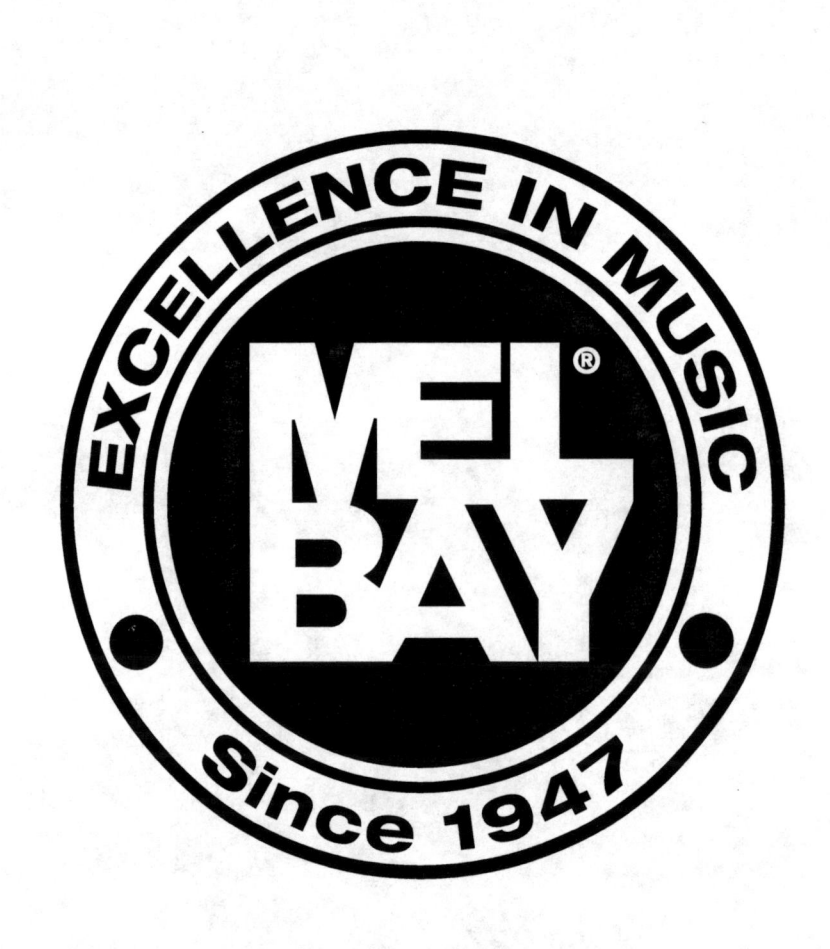